The Wonder[...] Solar System

Lisa E. Greathouse

Earth and Space Science Readers: The Wonder of Our Solar System

Publishing Credits

Editorial Director
Dona Herweck Rice

Associate Editor
Joshua BishopRoby

Editor-in-Chief
Sharon Coan, M.S.Ed.

Creative Director
Lee Aucoin

Illustration Manager
Timothy J. Bradley

Publisher
Rachelle Cracchiolo, M.S.Ed.

Science Contributor
Sally Ride Science

Science Consultants
Nancy McKeown
Planetary Geologist
William B. Rice,
Engineering Geologist

Teacher Created Materials Publishing
5301 Oceanus Drive
Huntington Beach, CA 92649
http://www.tcmpub.com

ISBN 978-0-7439-0561-9

Reprinted 2009

Table of Contents

Our Place in the Universe

Have you ever looked into the night sky and wondered what was out there? Throughout time, people have gazed to the heavens. They are hoping to find clues about our place in the universe.

Long ago, people thought that Earth was the center of the **solar system.** They thought that the sun, moon, stars, and other planets revolved around it. Scientists finally learned that Earth is one of at least eight planets in our solar system. They found that all of the planets travel around the sun.

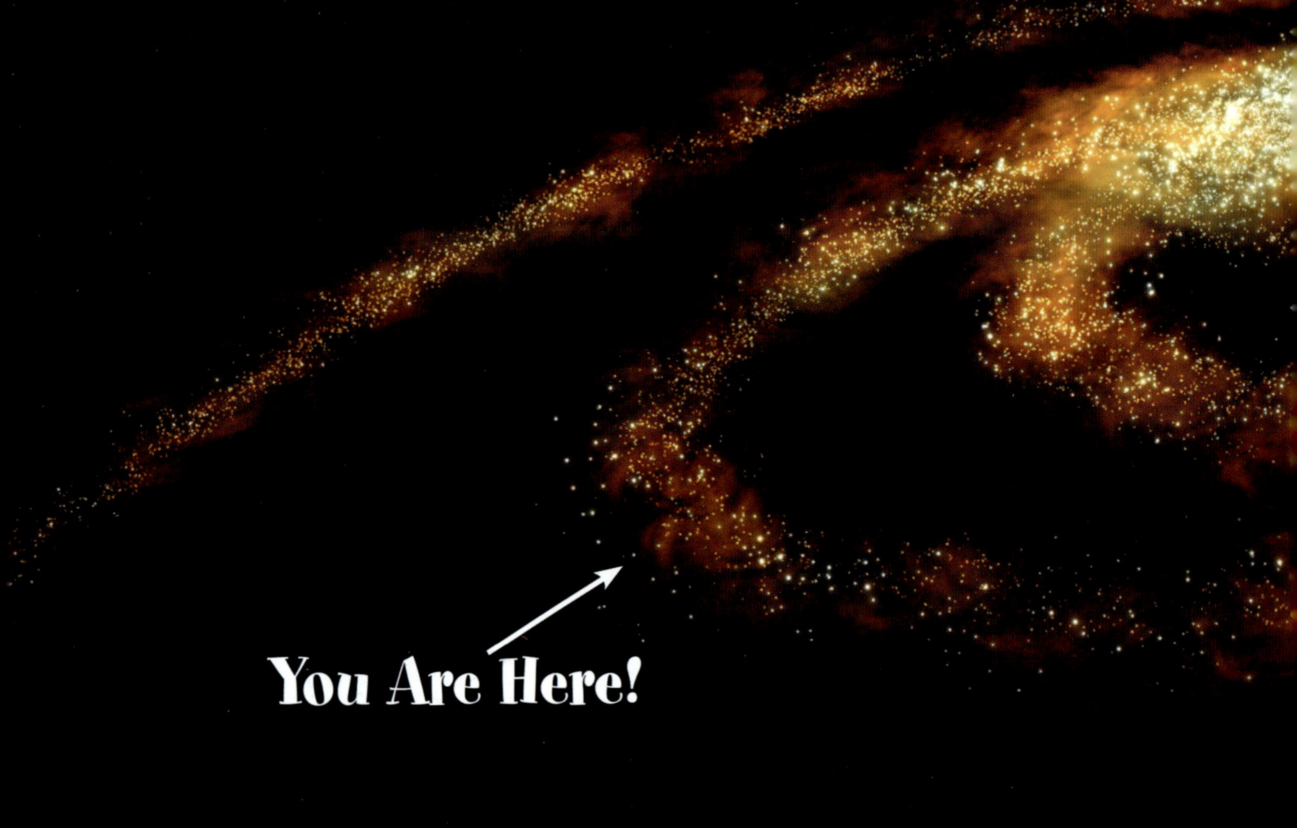

Our galaxy, the Milky Way, is a spiral galaxy.

Space Wanderers

The word planet comes from the Greek word for wanderer. Ancient Greeks noticed that some of the lights in the night sky seemed to wander. Now we know that planets move in an invisible path around the sun. That path is called an **orbit**.

Astronomers also now know that our solar system is almost 5 billion years old. It is located at the edge of the **Milky Way galaxy**. It is one of at least 100 billion galaxies in the universe. Each one of them has billions of stars! Our galaxy alone has about 200 billion stars.

The Milky Way is a spiral galaxy. It can be seen on a very dark night as a bright band in the sky. It is so wide that light would take 100,000 years to travel across it.

The Sun

The sun is a **star** at the center of the solar system. It isn't the biggest or brightest star in our galaxy, but it is the star closest to Earth. It is the largest body in our solar system. In fact, it contains 99.8 percent of all of the mass in our solar system.

The sun is the major source of energy on Earth. It gives us light and heat. It's responsible for the growth of plants, winds, ocean currents, and the water cycle. Without the sun, Earth would be very cold. It would be so cold that no living thing could survive.

Like other stars, the sun is made up of hot hydrogen and helium gases. The temperature in its center is about 15 million degrees Celsius (27 million degrees Fahrenheit).

The sun also creates **gravity**. The force of gravity between the sun and planets keeps Earth and other planets of the solar system in place and orbiting around the sun.

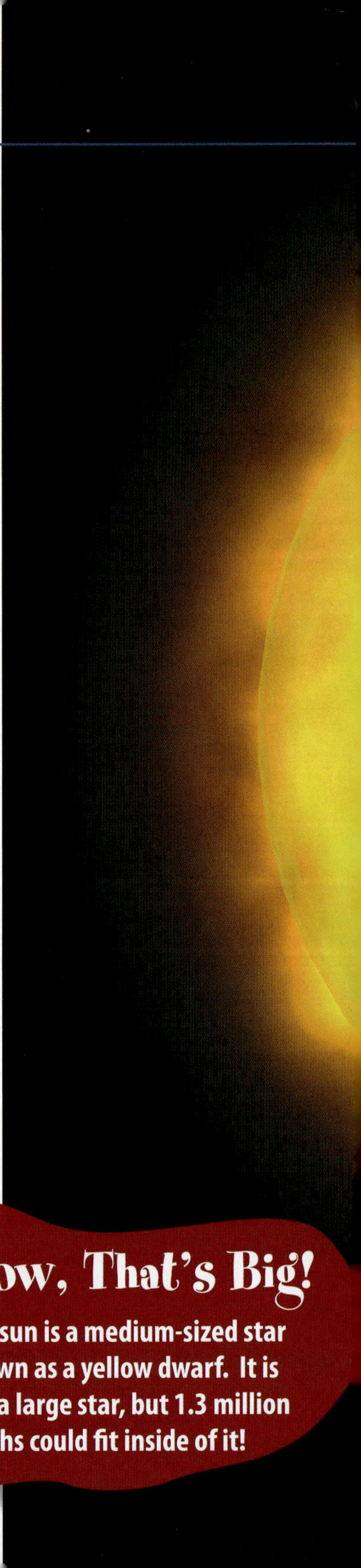

Now, That's Big!

The sun is a medium-sized star known as a yellow dwarf. It is not a large star, but 1.3 million Earths could fit inside of it!

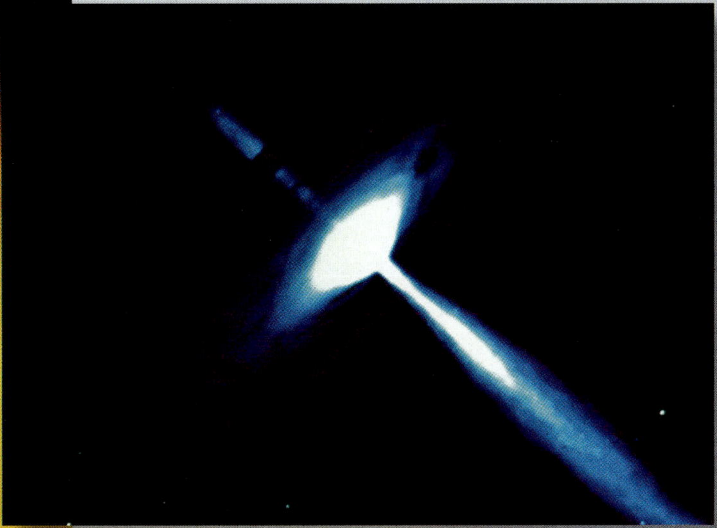

What Is a Black Hole?

Huge stars that use up all their energy form black holes. As the stars die, their gravity causes them to collapse. Their gravity is so great that not even light can escape. Any object that is pulled into a black hole becomes invisible. A black hole is not *really* a hole. It is an area very crowded with **matter**. We only know a black hole is there because of the way an object that gets near one is pulled toward it. One black hole that scientists found is only a few kilometers wide. But it weighs as much as three billion suns!

The Planets

My Very Educated Mother Just Served Us Nachos.

For years, people have used a sentence like the one above to remember the names of the eight planets. The first letter of each word stands for a planet. It also helps us remember the order of the planets from the sun. The planets are Mercury, Venus, Earth, Mars, Jupiter, Saturn, Uranus, and Neptune.

A **planet** is a large body that revolves in space around a star. It also reflects that star's light. The sun is the star around which the planets in our solar system revolve. Most of the planets have more than one **moon** traveling around them. While the planets are revolving around the sun, they're also spinning, or rotating. This is why we have day and night. When our part of Earth rotates away from the sun, it's night. The time it takes for a planet to rotate once is the planet's day. It takes Earth 24 hours to make one **rotation**, so Earth's day is 24 hours long.

Planet Races

The planets nearest the sun move the fastest. Mercury orbits the sun once every 88 days. It takes Neptune 165 years to travel once around the sun! A planet's year is the number of days it takes to orbit the sun.

What Would You Weigh on Another Planet?

Do you think you know how much you weigh? When you stand on a scale on Earth, your weight is really a measure of the pull of gravity between you and Earth. How strong that pull is on other planets depends on how big the planet is. On planets with more mass than Earth, you would feel heavier. On planets with less mass, you would feel lighter.

So, if you weigh 441 newtons (100 pounds) on Earth, you would weigh only 166 newtons (38 pounds) on Mercury. Mercury has a much smaller **mass** than Earth has. If you could stand on Jupiter, you would weigh 1048 newtons (236 pounds)! Jupiter is about the size of 318 Earths. That's a lot of mass. So, the person standing on it would weigh a whole lot, too.

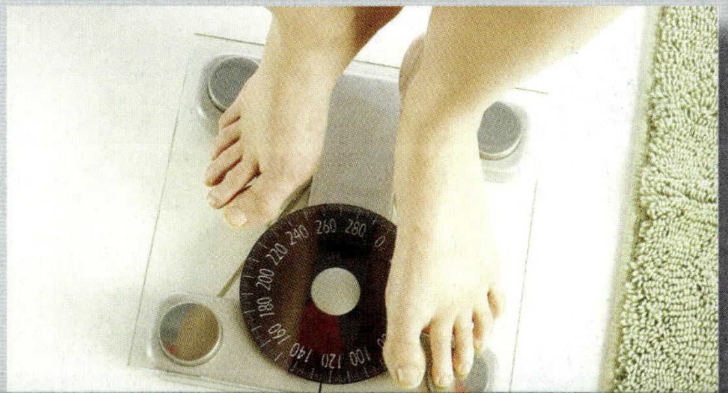

Mercury

Mercury and Venus

Mercury is the closest planet to the sun. It is called an **inner planet**. Venus, Earth, and Mars are inner planets, too. Mercury is the smallest of the four. All inner planets have rocky surfaces.

Mercury is covered with **craters**, or shallow holes. They were made when the planet was struck by other space objects. Mercury can get very hot. It can be hot enough to melt lead! Its high temperature is about 425 degrees Celsius (797 degrees Fahrenheit). But at night, it can get as cold as 150 degrees below zero Celsius (238 degrees below zero Fahrenheit). The big range is due to Mercury's long days. One day on Mercury is about the same as 59 Earth days. One night is about that long, too. It can be very cold away from the sun for 59 days, and it can be very hot so near to the sun!

Venus is the second planet from the sun. It's the hottest planet in our solar system. It's also the brightest planet in the night sky. Its thick clouds hold the sun's heat. It can get as hot as 462 degrees Celsius (864 degrees Fahrenheit)! Venus is slightly smaller than Earth. It spins much more slowly, though. It takes 243 Earth days for Venus to spin once. It takes 225 Earth days for Venus to orbit the sun. That means that Venus's day is 18 Earth days longer than its year!

Goddess Planet

Because Venus is the brightest planet in the solar system, the Romans named it after their goddess of love and beauty. Everything on the surface of Venus is named after goddesses and other women in history and literature.

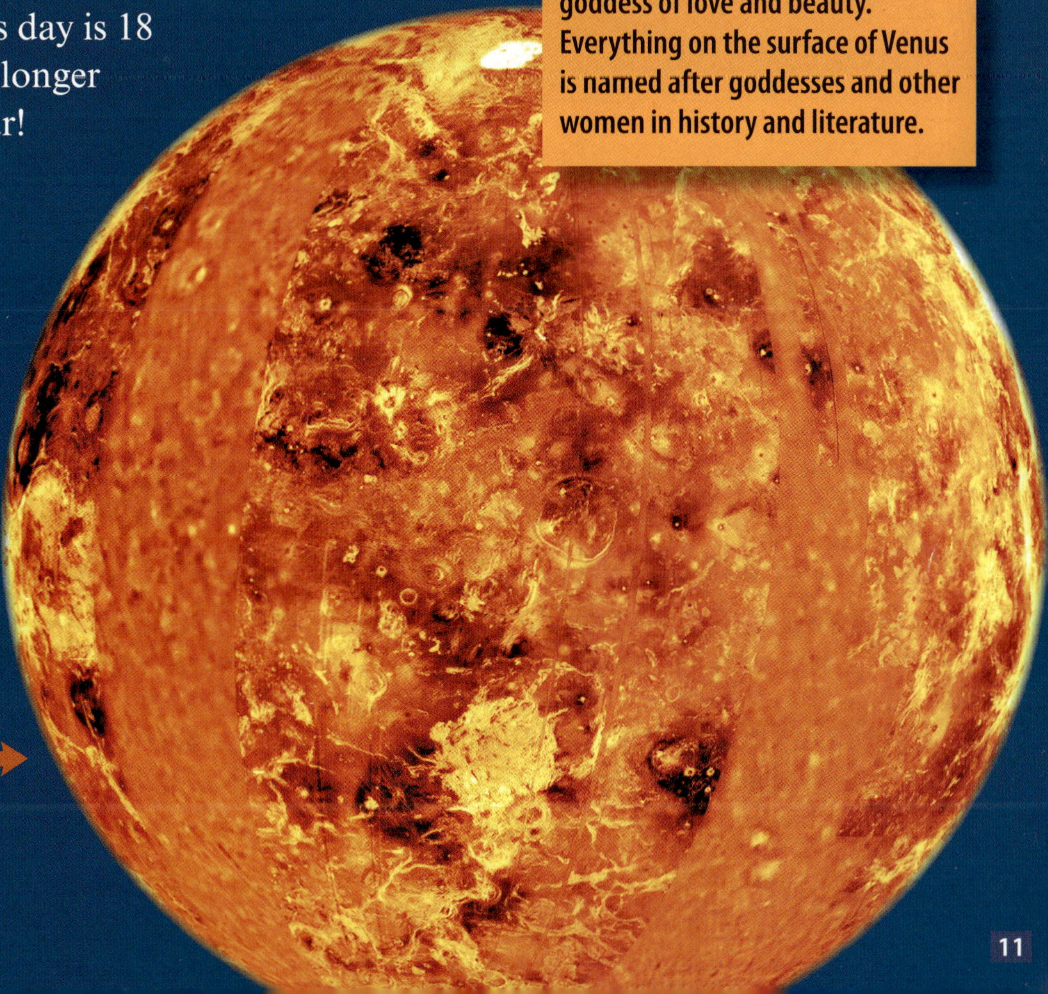

Venus

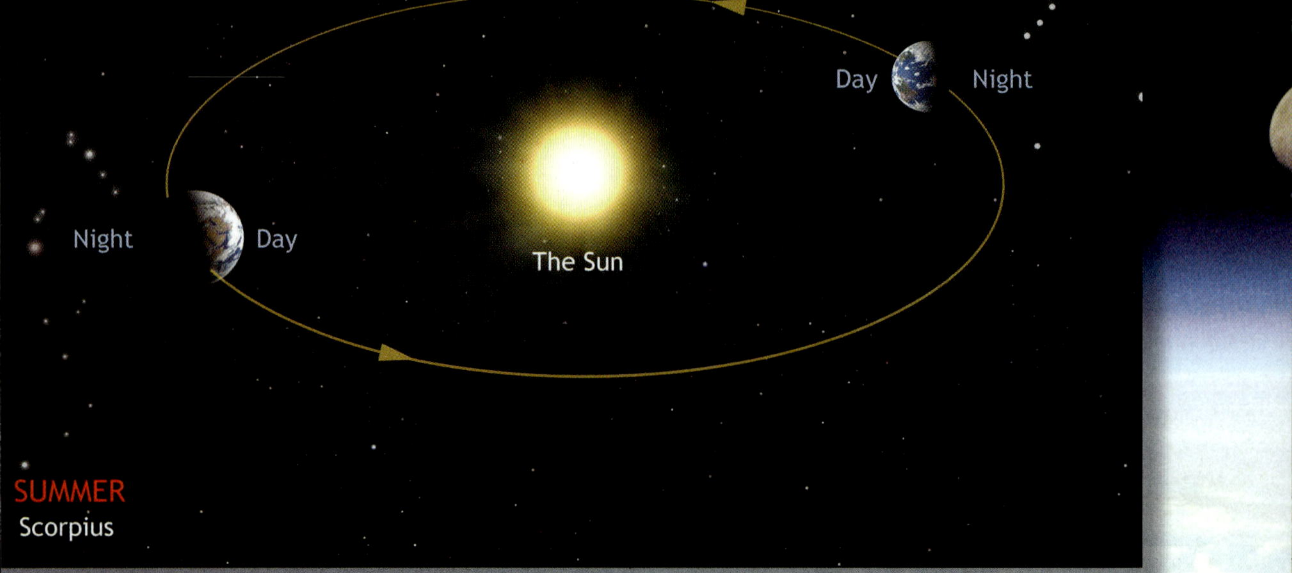

Earth completes one orbit around the sun each year.

Life on Earth

Earth is unlike any other planet in our solar system. It is the only planet known to have life. No other planet has oceans of water. Venus has rain, but it is highly acidic. We know that water is necessary for life as we know it. Water covers three-quarters of Earth's surface. That makes it appear blue from outer space. Earth's **atmosphere** contains oxygen that lets us breathe. It also protects us from the deadly rays of the sun.

Earth travels around the sun every 365 and $\frac{1}{4}$ days. That is one Earth year. As it orbits the sun, Earth is rotating at a speed of 1,660 kilometers (1,031 miles) per hour. In 24 hours, it makes one rotation. That is one Earth day.

Earth's Name

Earth is the only planet in our solar system that is not named after a god. Its name comes from the Old English word eorthe, meaning ground.

Earth's Moon

Only one moon orbits Earth. It is nearly as big as the planet Mercury!

This spectacular photo of Earth was taken from far into space. Parts of Africa, Europe, and Asia are visible.

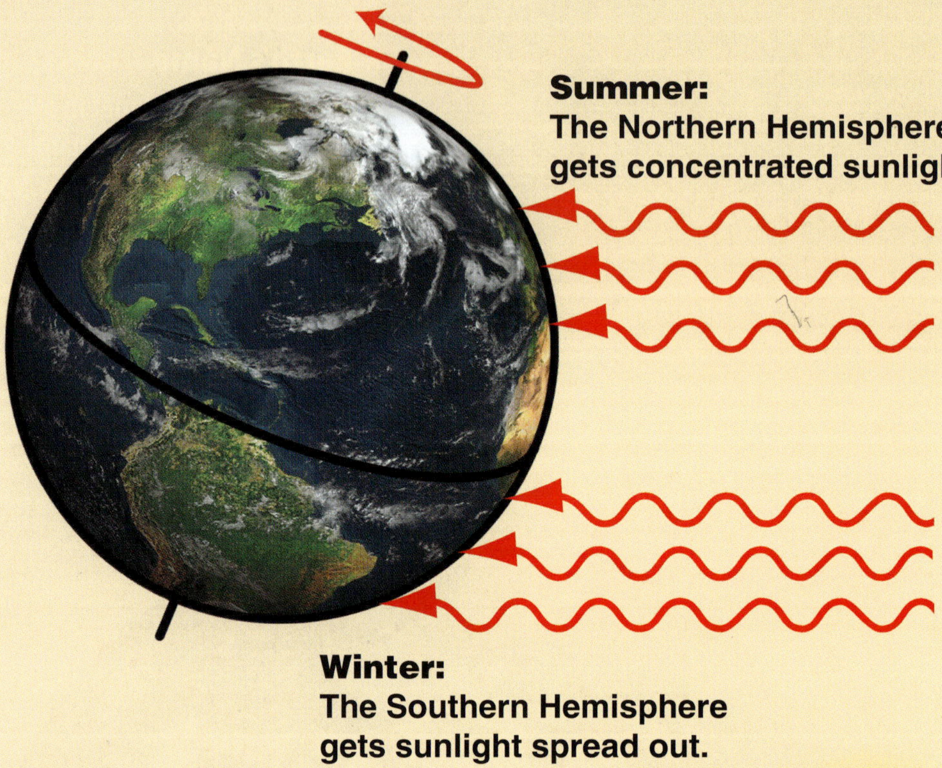

Four Seasons

The way Earth is tilted causes the seasons. For part of the year, the northern half of Earth leans toward the sun and gets direct sunlight. So, it's summer there. At the same time, the southern half leans away from the sun and gets less sunlight. So, it's winter there.

The tilt doesn't change. It is the position of the planet around the sun that changes, As Earth goes around the sun, different parts get more sunlight. That makes the seasons.

One Small Step

On July 29, 1969, over 600 million people watched their televisions. They saw astronaut Neil Armstrong become the first human to step on the moon. People met at moonwalk parties, restaurants, and wherever they could get near a television. Everyone wondered what the first words spoken on the moon would be. It seemed as if the whole world held its breath as Armstrong climbed down the ladder in his spacesuit. He stepped onto the moon's surface. Then he said these famous words. "That's one small step for man, one giant leap for mankind."

Mars and Jupiter

Mars is a cold, windy planet. It has the largest mountains in the solar system. Its largest volcano is more than three times higher than Earth's tallest mountain! Mars is called the Red Planet. That's because of its rusty color. Mars is about half the size of Earth. The average temperature is about 60 degrees below zero Celsius (140 degrees below zero Fahrenheit). Scientists think water may have flowed on Mars millions of years ago. They are still looking for signs that there may have been life there.

Jupiter is the largest planet in the solar system. In fact, it is more massive than the other seven planets combined! It is also one of four giant planets made of gases. Jupiter has at least 63 moons. It also has three faint rings. Its main feature is a huge storm called the Great Red Spot. This storm has lasted at least 300 years.

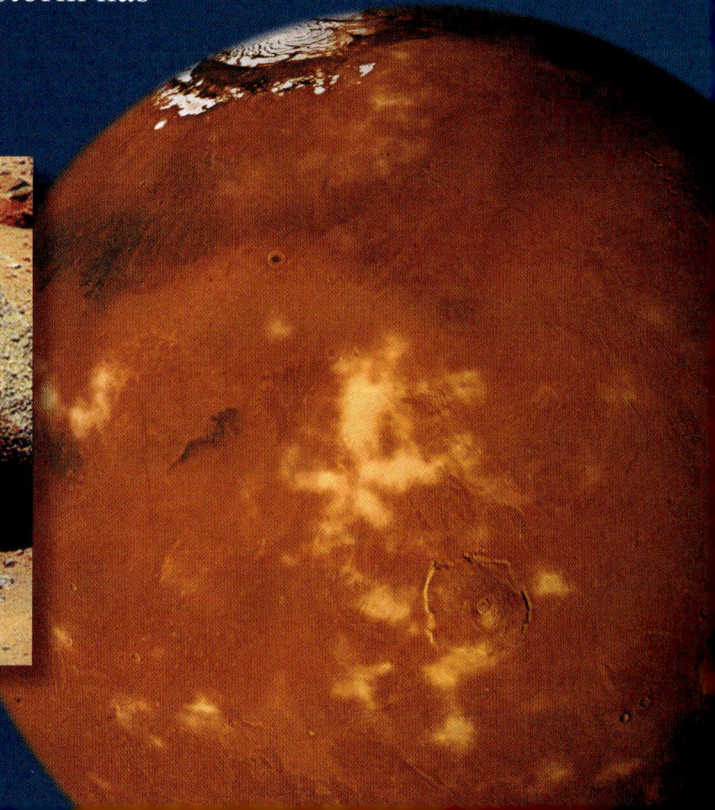

This reconstruction shows a rover exploring Mars.

Two rovers named *Spirit* and *Opportunity* have explored Mars for over four years!

Warriors and Kings

Mars is named for the Roman god of war. Its red color made people think of the blood of battle. Jupiter was named after the king of the Roman gods and goddesses. Its size made it the "king" of planets.

Jupiter's Great Red Spot

Mars

A Long Way to Go

Saturn is nearly twice as far from the sun as its neighbor, Jupiter. It takes almost 29 years to orbit the sun.

Saturn

Saturn and Uranus

Many people think **Saturn** is the most beautiful planet. That's because of its thousands of rings. These rings are made up of ice and rock. They are held in an orbit around the planet by their own speed and Saturn's gravity.

Saturn is the second largest planet in the solar system. It is made mostly of hydrogen and helium. It has at least 59 moons. The most famous one is Titan. It is the only moon in the solar system with an atmosphere.

Voyager 2 (arrow pointing up)

Uranus (arrow pointing left)

Uranus was the first planet discovered with a telescope. It looks like a blue-green disk lying on its side. Some astronomers think a crash with a planet-sized object may have knocked Uranus onto its side. *Voyager 2* was the only space probe to ever visit the planet. It helped discover many of its 27 known moons. It's believed that Uranus has 11 rings rotating around it.

two Neptunes

Neptune and Pluto

In 1846, scientists went looking for a planet they thought might be moving Uranus with its gravity. They charted where they thought this new planet might be. When they pointed their telescopes at that point, there was **Neptune**!

Neptune is the last of the four giant planets. It was named after the Roman god of the sea. That's because of its deep blue color. It has six rings and at least 13 moons. The weather on Neptune is fierce. Winds whip up to more than 1,995 kilometers (1,250 miles) per hour!

Scientists wanted to find a ninth planet. They tried the same trick as before. They charted where they thought it would be. When they looked, they found **Pluto**. It wasn't

discovered until 1930. It is over 5.9 billion kilometers (3.7 billion miles) away from the sun. It takes Pluto 248 years to orbit the sun just once.

Pluto is tiny. It is just two-thirds the size of our moon. Scientists argued whether it was big enough to be called a planet. They decided Pluto should be called a **dwarf planet**.

Hello, Pluto!

Spacecraft have traveled to the moon, the sun, and to all the planets except Pluto. An unmanned spacecraft is expected to reach the distant planet in 2015. The mission is called *New Horizons*.

Night and Day

Pluto is so far from the sun that there is almost no difference between night and day there.

Pluto

Dwarf Planets Galore!

Once scientists found Pluto, they started finding more specks in the sky everywhere. Planet-like objects orbiting the sun have been given names like *Sedna* and *Quaoar*. Some have been given codes like 2003 UB313 and 2004 DW.

Most of these objects have been found in the **Kuiper Belt**. That's a disk-shaped cloud of icy chunks beyond Neptune. There's also Ceres, which astronomers used to count as a planet until 1802. Then they decided it was just a big, round asteroid.

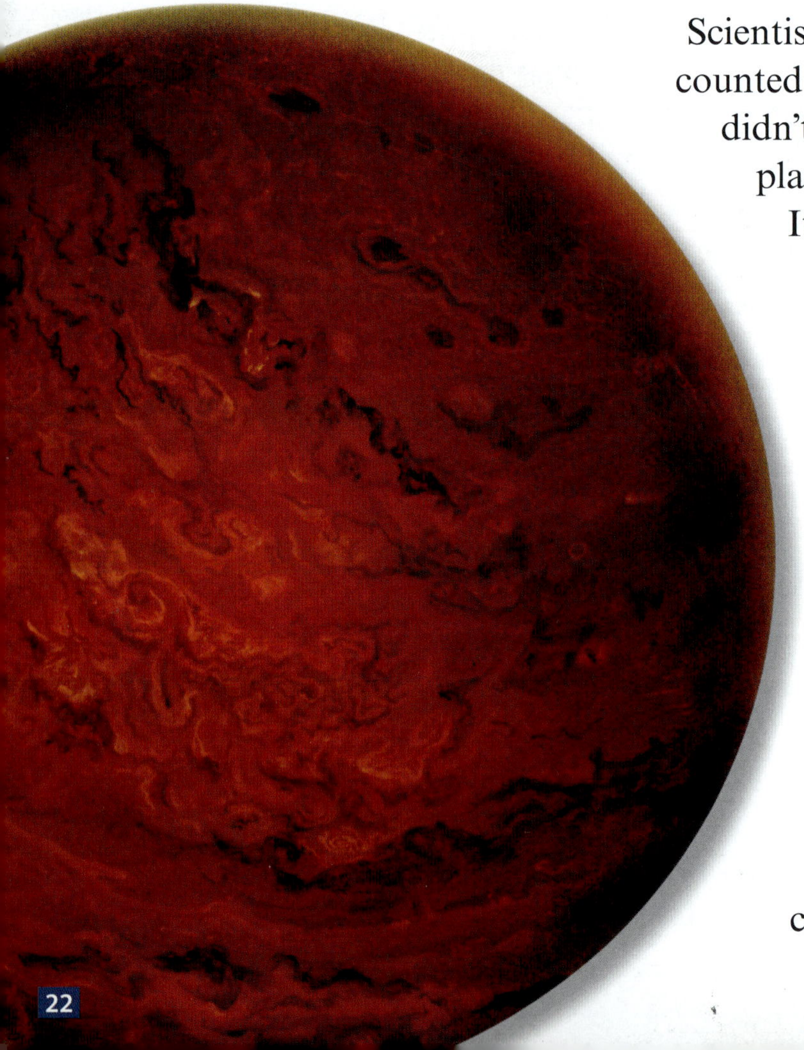

Scientists had to decide what counted as a planet and what didn't. They decided that a planet had to orbit the sun. It had to be round. It had to be the only thing in its orbit. Pluto's orbit crosses Neptune's orbit. It goes through the Kuiper Belt. There are all sorts of things in the way. Pluto can't be a planet.

Don't feel sorry for Pluto, though. It is joined by Ceres and Eris (or UB313). They are a brand new category called **dwarf planets**.

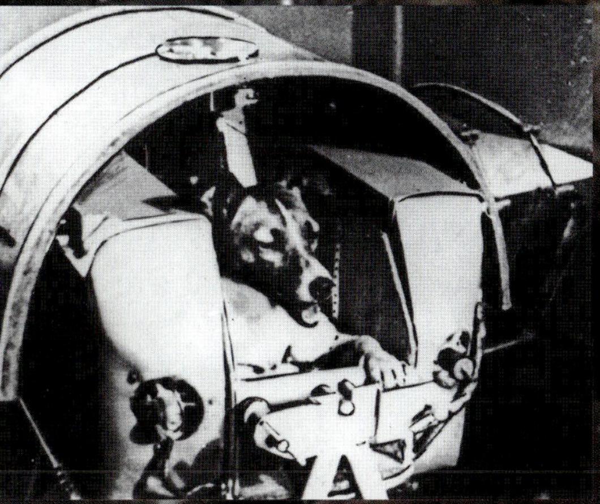

Animals in Space

It would not have been possible for humans to travel in space without animals going first. Scientists studied how animals reacted to conditions in space. This helped them prepare human astronauts. The first animal in space was Laika. She was a mixed breed dog sent by the Soviet Union in 1957. Her medical condition was watched from Earth. But she died in space. The United States has launched many animals into space. They include monkeys and mice. Many of them lost their lives. Without these animals, space travel would have been far more dangerous for humans.

Asteroids, Meteoroids, and Comets

There are many objects of all shapes and sizes in orbit around the sun. They were left over when the solar system was formed 4.6 billion years ago. Sometimes their orbits cross the orbits of planets.

Asteroids are chunks of rock. They can range in size from hundreds of kilometers across to just a few meters. Most are found in an asteroid belt between Mars and Jupiter.

Comets are balls of dust, rock, gas, and ice. They are sometimes referred to as "dirty snowballs." The snowball is the comet's head. When a comet gets close to the sun, its ice melts. Then we can see a bright streak of dust and gas millions of kilometers long. This is called the comet's tail. The dust and gas shine because of sunlight.

This image of the Hale-Bopp comet was taken on March 7, 1997. It is considered one of the brightest comets of all time.

Most **meteoroids** are very small rocks in space. They usually burn up while speeding through the atmosphere. Meteoroids that do this are known as **meteors**. They are also called shooting stars. Some large meteoroids have hit Earth's surface and made big craters. They are called **meteorites**.

When a meteorite hits the surface of a planet, the force of the impact carves out a crater.

Maria Mitchell

People have been curious about space for as long as people have been standing on Earth. But some people are more than just curious. They really try to figure out what is out there. The first woman in the United States to become an astronomer was Maria Mitchell. She also discovered a comet in the early 1900s. She paved the way for many more curious people to become astronomers, too!

Constellations

Centuries ago, humans used stars to help guide them. For example, the North Star is nearly above the North Pole. If people were facing that star, they knew they were facing north.

People also traced pictures in their minds around groups of stars. They traced animals, objects, or beings from myths. Names were given to 88 **constellations**. That gave people a way to chart the 6,000 stars seen with the naked eye.

Have you heard of Ursa Major? It is the constellation with the group of stars called the Big Dipper. Have you heard of Orion? People thought its row of three bright stars looked like the belt of a hunter from mythology.

Star Light, Star Bright

On a clear night away from city lights, you can see between 1,500 to 3,000 stars!

How High Can a Helium Balloon Float?

When you let a helium balloon go, do you wonder how high it can float? You can watch it for a while, but then you lose sight of it. Well, a helium balloon can fly very high, but it cannot go as high as outer space. As the balloon floats upward, the air around it gets thinner and colder. Because of these changes, the balloon gets larger the higher it travels. Eventually, the balloon gets so big that it bursts and the helium escapes into the atmosphere. How high the balloon goes depends upon how much helium it holds. Regular birthday balloons can rise as high as small airplanes fly!

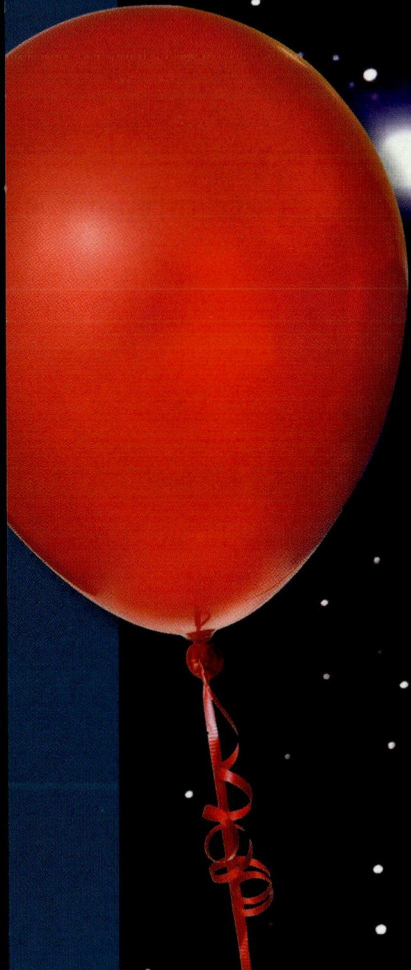

Lab: Collecting Micrometeorites

Did you know that items from outer space can be found in your own backyard? Tiny samples of space rocks float in our atmosphere. They float until dust and water carry them to the ground. These rock samples are micrometeorites. The best time to collect them is after a meteor shower.

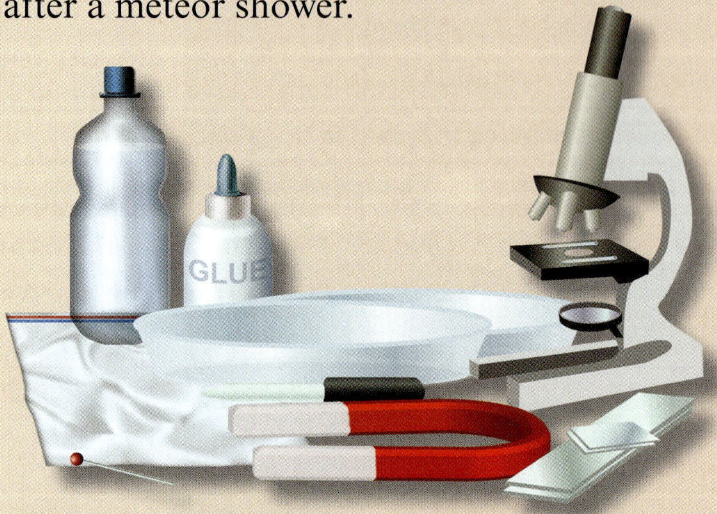

Materials

- 2 shallow dishes (heat resistant)
- distilled water (about 2 cups—enough to fill a dish)
- magnet
- plastic wrap or sandwich bag
- heat source for boiling water (optional)
- microscope or magnifying glass
- 2 microscope slides and covers
- mounting glue
- eyedropper
- sewing needle or large pin

Procedures

1 Place a dish outside to collect rainwater. Or leave the dish outside for a few days.

2 Cover a magnet with plastic wrap or a sandwich bag. Sweep the covered magnet through the water in the collecting dish, especially the bottom and sides. (Micrometeors are rich in iron, so they will stick to the magnet.)

3 Fill a second dish with distilled water. Place the covered magnet in the dish filled with distilled water. Remove the plastic from the magnet and gently swirl it around to allow the micrometeors to fall to the bottom of the dish.

4 Remove the plastic from the distilled water. Boil the water in the dish until it evaporates. (Ask an adult for help.) Or, let it evaporate naturally.

5 Magnetize a needle or pin by rubbing it on the magnet for about a minute. Drag the pin or needle along the sides and bottom of the dish.

6 Tap the needle or pin onto a microscope slide so samples fall onto the slide.

7 Glue down a cover glass. Examine the particles. Any rounded and pitted metallic particles are probably micrometeorites!

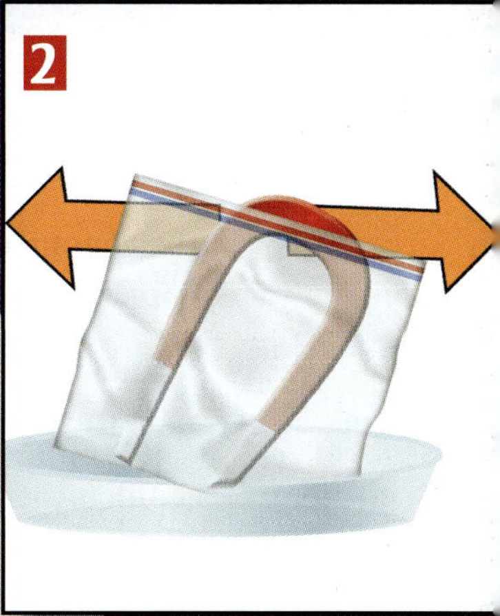

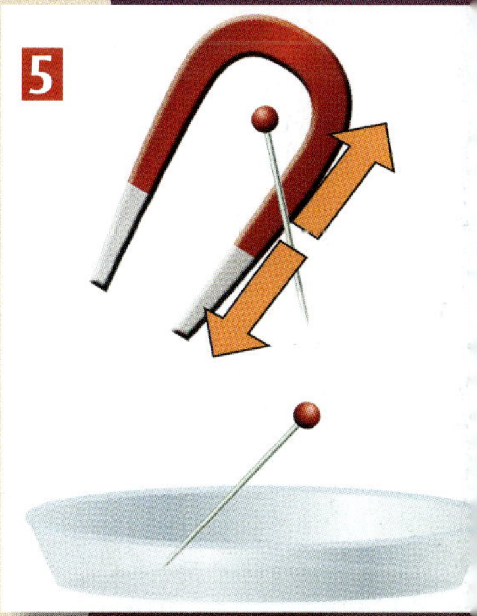

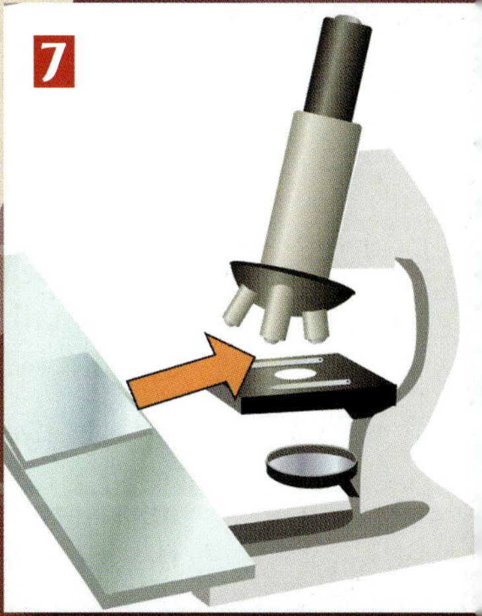

Glossary

asteroid—small body of rock that orbits the Sun

astronomer—a person who studies objects in space

atmosphere—a layer of gas that surrounds some planets and moons

comets—traveling objects made of ice, dust, and other material, that leave a glowing tail of gas and dust behind them when close to the sun

constellations—patterns formed by stars

crater—a bowl-shaped depression on the surface of a planet or moon caused by the impact of another body such as an asteroid or comet

dwarf planet—small, spherical mass in space that orbits a star but has not cleared its orbit of other objects

Earth—the planet third in order from the sun, on which human beings live

gravity—a force that pulls objects together

inner planets—the four small, rocky planets nearest the Sun

Jupiter—fifth planet from the sun, and the largest in the solar system

Kuiper Belt—a region of space filled with small, icy objects that orbit the sun far away

Mars—fourth planet from the sun

mass—the measure of the amount of matter in an object

matter—anything that occupies space and has mass

Mercury—planet closest to the sun and the smallest in the solar system

meteor—a streak of light produced by a meteoroid in Earth's atmosphere

meteorite—a piece of space rock that strikes a planet or moon

meteoroid—a solid object floating in space

Milky Way—the spiral galaxy in which our solar system exists

moons—natural satellites that orbit planets

Neptune—eighth planet from the sun

orbit—the path of an object as it moves around another object in space

planet—solid, spherical mass in space that orbits a star and has cleared its orbit of other objects

Pluto—dwarf planet whose orbit crosses Neptune

rotation—one complete spin of an object on its axis

Saturn—sixth planet from the sun

solar system—a group of planets that orbits a star

star—large luminous body in space that is made of gases

Uranus—seventh planet from the sun

Venus—second planet from the sun

Index

Sally Ride Science

Sally Ride Science™ is an innovative content company dedicated to fueling young people's interests in science. Our publications and programs provide opportunities for students and teachers to explore the captivating world of science—from astrobiology to zoology. We bring science to life and show young people that science is creative, collaborative, fascinating, and fun.

Image Credits

Cover: Jurgen Ziewe/Shutterstock; p.3 Jurgen Ziewe/Shutterstock; p.4 (top) Jurgen Ziewe/Shutterstock; p.4–5 NASA; p.5 NASA; p.6 (top) Shutterstock; p.6–7 Michael Lawlor/Shutterstock; p.7 Stock Trek/Getty Images; p.8 (top) NASA; p.8–9 Tim Bradley; p.9 Photos.com; p.10 (top) NASA; p.11 (top) Peter Arnold, Inc./Alamy; p.11 (bottom) Lars Lentz/iStockphoto; p.11 Galina Barskaya/Shutterstock; p.12 (top) Mark Garlick/Photo Researchers, Inc.; p.12–13 NASA; p.13 NASA; p.14 (left) Tim Bradley; p.14 (right) Tim Bradley; p.15 NASA; p.16 (left) Space Frontiers/Hulton Archive/Getty Images; p.16 (right) Lars Lentz/iStockphoto; p.17 (left) NASA; p.17 (right) NASA; p.17 (background) NASA; p.18 Photos.com; p.19 (left) NASA; p.19 (right) NASA; p.20 (left) NASA; p.20 (right) James Saunders/Shutterstock; p.21 (left) Detlev van Ravenswaay/Photo Researchers, Inc.; p.21 (right) Shutterstock; p.22 (top) Steve Gray/iStockphoto; p.22 (bottom) Phototake Inc./Alamy; p.23 (top) Keystone/Getty Images; p.23 (background) David A. Hardy/Photo Researchers, Inc.; p.24 (top) M.E. Mulder/Shutterstock; p.24 (bottom) Flickr; p.24–25 Detlev van Ravenswaay/Photo Researchers, Inc.; p.25 (right) The Granger Collection, New York; p.25 (left) Mark Garlick/Photo Researchers, Inc.; p.26 (top) William Attard McCarthy/Shutterstock; p.26 (left) Joe Tucciarone/Photo Researchers, Inc.; p.27 (left) Morgan Lane Photography/Shutterstock; p.27 (background) William Attard McCarthy/Shutterstock; p.28 (top) NASA; p.28–29 Nicoll Rager Fuller